Physics Wave Questions for Marine Engineering Applications

Dedication

Who is like you, LORD God Almighty?

You, LORD, are mighty, and your faithfulness surrounds you.

You rule over the surging sea;
 when its waves mount up, you still them.

(*Psalm 89 vs. 8-9* NIV)

Contents   *Page No.*

Dedication	2
Contents	3
Foreword	5
Questions	9
Worked Solutions	25
Equations and Key Constants	66
Other Titles by the Authors	98

Copyright © Christopher Lavers, Sara-Kate Lavers 2015
First Edition 2016 **ISBN 978-1-326-90539-2**

The Authors asserts the moral right to be identified as the author of this work. All rights reserved. No part of this publication may be reproduced, stored in a retrieval system, or transmitted, in any form or by any means, electronic, mechanical, photocopying, recording or otherwise, without the prior permission of the authors.

Foreword

This book is intended to develop the student's proficiency dealing with basic physics wave motion concepts underpinning Basic Electrotechnology found in most maritime-related physics and engineering courses, (including Naval, Coastguard, or Merchant Marine Engineering).

This book provides 50 questions with their corresponding fully worked solutions for a range of basic physics wave motion problems starting at a level consistent with those covered in Introductions: *Physics Wave Concepts for Marine Engineering Applications* (by Christopher Lavers and Sara-Kate Lavers) in the Reeds Marine Engineering and Technology Series ISBN: 9781472922151.

This books also provides a companion to *Physics Wave Concepts for Marine Engineering Applications*, and precedes *Further Physics Wave Concepts for Marine Engineering Applications*, bridging the gap to wave questions covered in *Basic Electrotechnology for Marine Engineers* (Volume 6 in the Reeds Marine Engineering and Technology Series (by Christopher Lavers), and *Advanced Electrotechnology for Marine Engineers* (Volume 7 in the Reeds Marine Engineering and Technology Series (by Christopher Lavers, and Edmund GR Kraal). *Further Physics Wave Concepts for Marine Engineering Applications* (in press Lulu.com) develops questions to a level commensurate with current Electrotechnology syllabi of the UK Department for Transport Examinations, and is suitable for study alongside both Reeds Volumes Volume 6 as well as other volumes of the Lavers' *Basic Topics for Marine Engineers series* (lulu.com).

Knowledge regarding Basic Wave motion principles and electromagnetic devices dependent upon these principles is now essential to merchant navy sea qualifying requirements, notably **Standards of Training, Certification and Watch keeping** (STCW95), as mandated by the UK Department for Transport Maritime Coastguard Agency (MCA). This particular questions and answers volume was designed for self-study by Marine Engineers who may be at sea, and physicists, and was written in as simple a manner as possible to support increasing number of students for whom English is not their first language of study. These questions will support your own self-study and learning. It is very important that you consider setting and solving questions at least as difficult as those found here on which to practice your increasing skills of numeracy.

Christopher Lavers, Sara-Kate Lavers

Physics Wave Questions for Marine Engineering Applications

Q1. Find the wavelength of the following electromagnetic waves:

a. 10 GHz radar, (2 decimal places)
b. 3 MHz radio, and (1 significant figure), and
c. 10^{13} Hz microwaves, (1 significant figure)

Q2. Find the time delay t in terms of the Periodic Time T for a wave which is 60° out-of-phase with another identical wave (2 decimal places).

Q3. The muzzle of a starting pistol at a race is seen to flash. If the sound of the gun is heard 0.5 seconds later, how far away is the gun (1 decimal place)? Take the speed of sound to be 330 metres per second (at sea level).

Q4. Find the frequency for a sound wave in water travelling at a speed of 1510 metres per second and having a wavelength of 10 cm (3 significant figures).

Q5. In the United Kingdom 'mains frequency' is 50 Hz. What is the corresponding wave period (2 decimal places)?

Q6. If the group velocity is given by $V_g = \frac{\omega}{k}$ and the phase velocity is given by $V_p = \frac{d\omega}{dk}$ show how the phase velocity can be written in terms of the group velocity.

Q7. If $V_g = V_0 e^{-2kx-\omega t}$ find V_p in terms of V_g.

Q8. If $y(t) = \cos(\omega t)$ find by differentiation the maximum and minimum values of $y(t)$, where ω is the angular velocity.

Q9. If $y(t) = A\cos(\omega t)$, the velocity is 6 ms^{-1} and the acceleration 3 ms^{-2} find the first solution for the angular velocity ω multiplied by time t (ωt).

Q10. For a pendulum where $\omega = \sqrt{\frac{g}{l}}$ and l is the length of the pendulum if f = 0.5 Hz and $\omega = 2\pi f$ find the length of the pendulum (2 decimal places).

Q11. For waves in deep water find the phase velocity for 20 m wavelength gravity waves (2 decimal places) using the equation $c = \sqrt{\frac{g\lambda}{2\pi}}$.

Q12. For shallow water waves of wavelength 5 m find the phase velocity (1 decimal place).

Q13. Deep water wave amplitude is seen to decrease with depth according to the equation $A = A_0 \exp\left(-\frac{2\pi z}{\lambda}\right)$, where $A_0 = \frac{\lambda}{2\pi}$. Find the value A when $\lambda = 0.5\ m$ and z = 8 m (2 decimal places).

Q14. Find the gradient of the curve (dA/dz) in Q13.

Q15. Using the expression: $\frac{p}{\rho} + \frac{v^2}{2} + gH = constant$ find the final pressure p_{Final} if the final velocity v_{Final} is twice the initial velocity $v_{Initial}$ and the final height H_{Final} is twice the initial height $H_{Initial}$.

Q16. The following list shows 5 different wave motions:

> 30 MHz radio signal,
> 10 GHz radar,
> Ultra-Violet light,
> 10 kHz sonar, and
> 1 kHz sound in air.

a. Which has the longest wavelength?
b. Which has the next highest wavelength after Ultra-Violet radiation?

Q17. Place the following in order of frequency, starting with the shortest:

Red light, 10 GHz radar, 6 micron infra-red radiation, 5 kHz sonar, 3 kHz sound in air, and 3000 m radio.

Q18. For a single photon of wavelength 500 nm, what is the energy of the radiation emitted (3 significant figures)?

Q19. For the photon of question 18 what is the energy of the photon in electron Volts (3 decimal places)?

Q20. Consider an X-ray source. What will be the thickness of absorber required for the incident radiation level to be reduced to 1/6 of its initial value in terms of α if $\alpha = 0.5$ m^{-1} (3 decimal places)?

Q21. Use differentiation and the equation $I_x = I_0 e^{-\alpha x}$ to show the rate of change of intensity with distance.

Q22. If the rate of change of intensity at a distance x is -6 Wm^{-1} and the intensity at distance x is 3.7 W, what is the decay constant α in m^{-1} (2 decimal places)?

Q23. If the individual photon packet of an X-ray has a frequency of 1.7×10^{19} Hz and the overall intensity of the X-rays is recorded as 1.5×10^9 Wm^{-2} across an area of 1 square metre in 1 second what will be the total number of X-ray photons produced with constant flux (2 significant figures)?

Q24. Find the critical angle for a light wave travelling from diamond (refractive index = 2.4), into water having refractive index = 1.333 (2 decimal places).

Q25. Use the following equation to calculate the missing variable in each situation:

λ = dx/L where λ is the wavelength of the light, x is the fringe spacing, d is the distance between slits, and L is the distance from the slits to the screen.

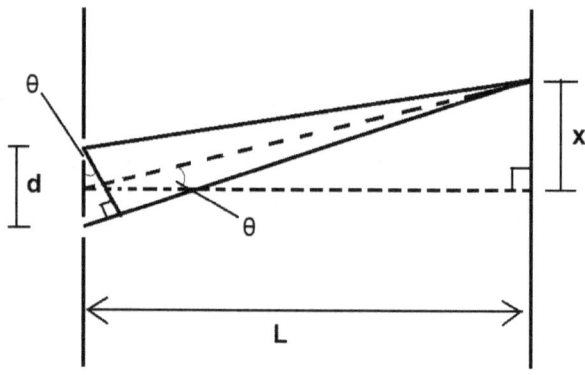

i) The wavelength of the light in the instance where x = 0.04 cm, d = 2.5 mm and L = 2.6 m (nearest whole nm).

ii) What distance must the screen be from the slits for the fringe spacing to be twice the distance between the slits for ultra-violet light when x = 1 mm and L = 2 m?

Q26. Consider a wave incident at a boundary between air and water. The air has a refractive index of 1.0003, whilst the water has a refractive index of 1.333. The wave is incident at an angle of 60 degrees to the normal. Find the transmitted angle θ_t in the water, and sin θ_t (both 3 decimal places).

Q27. Consider a wave incident at a boundary between water and glass. The air has a refractive index of 1.333, whilst the glass has a refractive index of 1.52. The wave is incident at an angle of 10 degrees to the normal. Find the transmitted angle θ_t in the water, and sin θ_t (both 3 decimal places).

Q28. A radar transmitter has a 15 W power source, if the source radiates isotropically (equally) in all directions what will the intensity be 20 m from the source (3 significant figures)?

Q29. A radar set detects a target at a range of 55 km. The intensity of the signal striking the target is 13×10^{-2} W m^{-2}.

Calculate the new intensity if the range is decreased to 25 km and the transmitted power is increased by 40 % (3 decimal places)?

Q30. For a target at a range of 20 km and the same target then recorded at a range of 65 km what will be the ratio of the detected echo strengths (5 decimal places)?

Q31. A radar set can just detect a target at a maximum range of 100 km. If the power output of the transmitter is increased by 35 % what will be the new maximum range of the radar in km (4 significant figures)?

Q32. What is the ratio of scattering intensity for a wavelength of 550 nm compared with a wavelength of 750 nm (4 decimal places)?

Q33. Using the beam attenuation coefficient $C(\lambda)$ = 0.096 at 575 nm and for a water depth of 10 m and incident light intensity I_{in} = 4 Wm^{-2} find the outgoing light intensity I_{out} (2 decimal places). Use the equation: $I_{out} = I_{in} e^{-C(\lambda)x}$.

Q34. For water at a 470 nm wavelength, with C = 0.02 m^{-1} a water depth of 7 m, and surface incident light intensity I_{in} = 2.6 Wm^{-2}, what is the light intensity I_{out} recorded at the sea bed (2 decimal places)?

Q35. The vertical and horizontal beam widths of a radar antenna are 60° and 2° respectively. The radar operates at 3.0 GHz. Calculate the vertical and horizontal dimensions of the antenna (2 decimal places).

Q36. The vertical and horizontal dimensions of a radar antenna are 15 cm and 5.2 m respectively. If the radar operates at 3.2 GHz, what are the vertical and horizontal beam widths (2 decimal places)?

Q37. A diffraction grating of 4000 lines per centimetre is used with white light. How many orders of spectra will be observed?

Q38. With two radio receivers calculate the resolution limit if the wavelength is 2 m and the two receivers are spaced 2 km apart (2 decimal places).

Q39. With two receivers calculate the resolution limit in space if the wavelength is 1 m and there are now 30 receivers spaced 4 km apart (4 significant figures).

Q40. What is the receiver spacing in metres if a linear array can resolve a beam width of $0.02°$? The wavelength is 2 cm with 30 elements in a *sonar* array (2 decimal places).

Q41. If a receiver R at a distance x from both sources A and B is moved a distance *dx* towards source A calculate the distance needed to move from the central maximum to the **second** destructive minimum.

Q42. For two waves of the same frequency emitting with a time delay of T/3 and a physical path distance of $\lambda/4$ and a phase difference of 180 degrees what will be the resultant of the two waves arriving at the same point in space?

Q43. Find the time delay required in terms of the periodic time T if the path difference between two adjacent sources is 3 cm and the wavelength is 5 cm.

Q44. Consider a regularly spaced square array of sources in both the vertical and horizontal direction. If there are 1600 sources in total and the sources in every row and column are spaced by half a wavelength what will be the minimum resulting beam width if the wavelength is 1 cm (3 decimal places)?

Q45. Find the time delay required to steer a radar beam off the bore sight (the normal straight through direction) if the required path difference between adjacent sources is 3 cm (1 significant figure).

Q46. If the transmitted radar frequency is 3 GHz and the relative velocity is 20 metres per second and the speed of light c is taken as usual find the one way frequency shift observed by a distant observer (1 decimal place).

Q47. In the case of a Doppler shifted echo, calculate the velocity of the target if the observer is stationary, the transmitted frequency is 8.9 GHz and the echo shift is 4 kHz (3 significant figures).

Q48. Use: $V_{relative} = V_{source} \cos \theta_{source} - V_{target} \cos \theta_{target}$ to calculate the echo Doppler shift in wavelength if the source has a value of 50 metres per second on a bearing of 60 degrees and the target has a value of 15 metres per second on a bearing of 130 degrees and the radar is transmitting on a frequency of 9.5 GHz (2 decimal places).

Q49. An amplifier has a power gain of + 15 dB. If an input power is 13 mW dB, what is the output signal power (3 significant figures)?

Q50. If the initial power level P_1 is 5 W and the final power level P_2 is 100 W what is the Power ratio in decibels (2 decimal places)?

Physics Wave Worked Solutions for Marine Engineering Applications

A1. Find the wavelength of the following electromagnetic waves:

  a. 10 GHz radar, (2 decimal places)
  b. 3 MHz radio, and (1 significant figure)
  c. 10^{13} Hz microwaves, (1 significant figure)

Using V = fλ and rearranging the equation for the wavelength.

λ = V/f and substituting for the values given.
a. so $\lambda = 3 \times 10^8 / (10 \times 10^9) = 0.03$ m
b. so $\lambda = 3 \times 10^8 / (3 \times 10^6) = 100$ m
c. so $\lambda = 3 \times 10^8 / (1 \times 10^{13}) = 3 \times 10^{-5}$ m

A2. Find the time delay t in terms of the Periodic Time T for a wave which is 60° out-of-phase with another identical wave (2 decimal places).

Using the equivalence relationship:
$$\frac{t}{T} = \frac{\phi}{360}$$

So $t = \frac{\phi}{360} \times T = \frac{60}{360} \times T = 0.17\,T$ seconds

A3. The muzzle of a starting pistol at a race is seen to flash. If the sound of the gun is heard 0.5 seconds later, how far away is the gun (1 decimal place)? Take the speed of sound to be 330 metres per second (at sea level).

$$\text{Consider speed } v = \frac{\text{distance } d}{\text{time t taken to cover this distance}}$$

The sound of the gun arrives 0.5 seconds after the gun is fired.

From our speed equation above $t_1 = \frac{x}{v}$

Thus $0.5 = \frac{x}{v}$ and hence: $x = 0.5v = 0.5 \times 1500 = 750\ m$

A4. Find the frequency for a sound wave in water travelling at a speed of 1510 metres per second and having a wavelength of 10 cm (3 significant figures).

Using $V = f\lambda$ and rearranging the equation for the frequency.

$$f = V/\lambda = \frac{1510}{0.1} = 15100\ Hz \text{ or } 15.1\ kHz$$

A5. In the United Kingdom 'mains frequency' is 60 Hz. What is the corresponding wave period (2 decimal places)?

Using $f = \frac{1}{T}$ and rearranging for the period T:

Using $T = \frac{1}{f} = \frac{1}{60} = 0.02$ seconds

A6. If the group velocity is given by $V_g = \frac{\omega}{k}$ and the phase velocity is given by $V_p = \frac{d\omega}{dk}$ show how the phase velocity can be written in terms of the group velocity.

Generally $V = f\lambda$ so $\frac{dV}{dt} = f \times \frac{d\lambda}{dt} + \frac{df}{dt} \times \lambda$ if both variables are changing with respect to time.

$$V_g = \frac{\omega}{k} \quad V_p = \frac{d\omega}{dk}$$

$$V_p = \frac{d\omega}{dk} = \frac{d(V_g k)}{dk} = \frac{V_g dk}{dk} + \frac{k dV_g}{dk}$$

$$V_g = \frac{\omega}{k}$$

Thus: $\frac{dV_g}{dk} = -\omega^2/k$

And so:

$$V_p = \frac{V_g dk}{dk} + \frac{k dV_g}{dk} = V_g - k\omega/k^2 = V_g - \omega/k$$

A7. If $V_g = V_0 e^{-2kx-\omega t}$ find V_p in terms of V_g.

From the previous question we found that:

$$V_p = \frac{V_g dk}{dk} + \frac{k dV_g}{dk} = V_g - k\omega/k^2 = V_g - \omega/k$$

Since $V_g = V_0 e^{-2kx-\omega t}$ and using $V_p = V_g + \frac{k dV_g}{dk}$

Then: $V_p = V_g = V_0 e^{-2kx-\omega t} + \frac{k d(V_0 e^{-2kx-\omega t})}{dk} =$

$$V_0 e^{-2kx-\omega t} + k(-2x) V_0 e^{-2kx-\omega t}$$

$= V_0 e^{-2kx-\omega t}(1 - 2kx)$
$= V_g (1 - 2kx)$

A8. If $y(t) = \cos(\omega t)$ find by differentiation the maximum and minimum values of $y(t)$ where ω is the angular velocity.

$$y(t) = \cos(\omega t)$$

$\dfrac{dy(t)}{dt} = -\omega \sin(\omega t) = 0$ for maximum or minimum.

Thus either $\omega = 0$ as a trivial solution or $\sin(\omega t) = 0$ from which: $\omega t = \sin^{-1}(0) = 0$ degrees or $\omega t = \pi, 2\pi$ etc.

To find whether this is a maximum or minimum we need to look at the second derivative or rather: $a(t) = \dfrac{d^2 y}{dt^2}$

Now: $a(t) = \dfrac{d^2 y(t)}{dt^2} = -A\omega^2 \cos(\omega t)$

for a maximum or minimum to occur and will be +ve for minimum and –ve for maximum.

In the case above $a(t) = \frac{d^2y(t)}{dt^2} = -sin(90) = -1$ which is negative making this a maximum, which of course can be proven by inspection of the cosine function.

A9. If $y(t) = \cos(\omega t)$, the velocity is 6 ms^{-1} and the acceleration 3 ms^{-2} find the first solution for the angular velocity ω multiplied by time t (ωt).

$$y(t) = \cos(\omega t)$$

$$v(t) = \frac{dy(t)}{dt} = -\omega \sin(\omega t) \quad \text{and,}$$

$$a(t) = \frac{d^2 y(t)}{dt^2} = -\omega^2 \cos(\omega t)$$

So:

$6 = -\omega \sin(\omega t)$ (9.1) and
$3 = -\omega^2 \cos(\omega t)$ (9.2)
From equation (9.1) and rearranging:

$$\omega = \frac{-6}{\sin(\omega t)} \quad (9.3)$$

and from substitution into equation (9.2)

$$3 = -\frac{36}{\sin^2(\omega t)} \cos(\omega t)$$

And simplifying:

$$3\sin^2(\omega t) = -36\cos(\omega t)$$

$$3(1 - \cos^2(\omega t)) = -36\cos(\omega t)$$

$$3\cos^2(\omega t) - 36\cos(\omega t) - 3 = 0 \text{ or}$$

$$3\cos^2(\omega t) - 36\cos(\omega t) - 3 = 0$$

Which simplifies further to:

$$\cos^2(\omega t) - 12\cos(\omega t) - 1 = 0$$

$$\cos(\omega t) = \frac{+36 \pm \sqrt{36 \times 36 - 4 \times 3 \times (-3)}}{2 \times 3}$$

So

$$\cos(\omega t) = -0.00827625$$

$$\omega t = \cos^{-1}(-0.00827625)$$

$$\omega t = 94.747373 \text{ degrees.}$$

A10. For a pendulum where $\omega = \sqrt{\frac{g}{l}}$ and l is the length of the pendulum if f = 0.5 Hz and $\omega = 2\pi f$ find the length of the pendulum (2 decimal places).

For a pendulum $\omega = \sqrt{\frac{g}{l}}$ $\quad \omega = 2\pi f$ so $\omega = \sqrt{\frac{g}{l}} = 2\pi \times 0.5$

Thus $\frac{g}{l} = (4\pi)^2 \times (0.5)^2 = 39.478416$ and so: $l = \frac{g}{(4\pi)^2 \times (0.5)^2} = 0.25\ m$

A11. For waves in deep water find the phase velocity for 20 m wavelength gravity waves (2 decimal places) using the equation $c = \sqrt{\frac{g\lambda}{2\pi}}$.

$c = \sqrt{\frac{g\lambda}{2\pi}} = \sqrt{\frac{9.81 \times 20}{2\pi}} = 5.59$ metres per second.

A12. For shallow water waves of wavelength 5 m find the phase velocity (1 decimal place). In shallow water:

$$c = \sqrt{gz} = \sqrt{9.81 \times 5} = 7.0 \text{ metres per second.}$$

A13. Deep water wave amplitude is seen to decrease with depth according to the equation $A = A_0 \exp\left(-\frac{2\pi z}{\lambda}\right)$,

where $A_0 = \frac{\lambda}{2\pi}$. Find the value A when $\lambda = 0.5\ m$ and z = 8 m (2 decimal places).

$A = \frac{\lambda}{2\pi} \exp\left(-\frac{2\pi z}{\lambda}\right)$ substituting for the values given: $A = \frac{0.5}{2\pi} \exp\left(-\frac{2\pi 8}{0.5}\right) =$

$$A = \frac{1}{4\pi} \exp(-32\pi)$$

= 1.75 x 10^{-45} m, essentially zero.

A14. Find the gradient of the curve (dA/dz) in Q13.

$$A = \frac{\lambda}{2\pi} \exp\left(-\frac{2\pi z}{\lambda}\right)$$

The gradient of the curve is given by: $\frac{dA}{dz}$

so:

$$\frac{dA}{dz} = -\frac{2\pi\lambda}{2\pi\lambda} \exp\left(-\frac{2\pi z}{\lambda}\right) = -\frac{2\pi}{\lambda} A = -kA$$

where $k = 2\pi/\lambda$.

A15. Using the expression: $\frac{p}{\rho} + \frac{v^2}{2} + gH = $ constant find the final pressure p_{Final} if the final velocity v_{Final} is twice the initial velocity $v_{Initial}$ and the final height H_{Final} is twice the initial height $H_{Initial}$.

Since: $\frac{p_{Initial}}{\rho} + \frac{v_{Initial}^2}{2} + gH_{Initial} = \frac{p_{Final}}{\rho} + \frac{v_{Final}^2}{2} + gH_{Final}$ Then: $\frac{p_{Initial}}{\rho} + \frac{v_{Initial}^2}{2} + gH_{Initial} = \frac{p_{Final}}{\rho} + \frac{4v_{Initial}^2}{2} + 2gH_{Initial}$

$\frac{p_{Final}}{\rho} = \frac{p_{Initial}}{\rho} + \frac{v_{Initial}^2}{2} - \frac{4v_{Initial}^2}{2} + gH_{Initial} - 2gH_{Initial}$

Rearranging for final pressure: $p_{Final} = \rho(-gH_{Initial} - \frac{3v_{Initial}^2}{2} + \frac{p_{Initial}}{\rho})$

A16. The following list shows 5 different wave motions:

    30 MHz radio signal
10 GHz radar
    Ultra-Violet light
    10 kHz sonar, and
    1 kHz sound in air.

a. Which has the longest wavelength?

b. Which has the next highest wavelength after Ultra-Violet radiation?

The following list shows 5 different wave motions:

3 GHz radar $\lambda = \dfrac{3 \times 10^8}{10 \times 10^9} = 0.03 \ m$

Ultra-Violet light typically 200-400 nm

30 MHz radio signal $\lambda = \dfrac{3 \times 10^8}{30 \times 10^6} = 10 \ m$

10 kHz sonar, and $\lambda = \dfrac{1500}{10 \times 10^3} = 0.15 \ m$

1000 sound in air $\lambda = \dfrac{330}{1000} = 0.33 \ m$

    a. 30 MHz radio.
    b. 10 GHz radar.

A17. Place the following in order of frequency, starting with the shortest:

Red Light, 10 GHz radar, 6 micron infra-red radiation, 5 kHz sonar, 3 kHz sound in air, and 3000 m radio.

Placed in order of frequency, starting with the lowest:

(4) 10 GHz radar,

(5) 6 micron infra-red radiation $f = \frac{3 \times 10^8}{6 \times 10^{-6}} = 5 \times 10^{13}\ Hz$,

(2) 5 kHz sonar,

(1) 3 kHz sound in air,

(6) Red light $= \frac{3 \times 10^8}{700 \times 10^{-9}} = 4.3 \times 10^{14}\ Hz$

and,

(3) 3000 m radio $= \frac{3 \times 10^8}{3000} = 100\ kHz$.

A18. For a single photon of wavelength 500 nm, what is the energy of the radiation emitted (3 significant figures)?

$$f = c/\lambda = \frac{3 \times 10^8}{500 \times 10^{-9}} = 6 \times 10^{14} \, Hz$$

and

$$E = hf = 6.62 \times 10^{-34} \times 6 \times 10^{14}$$
$$= 3.97 \times 10^{-19} J$$

A19. For the photon of question 18 what is the energy of the photon in electron Volts (3 decimal places)?

If $1 \, eV = 1.6 \times 10^{-19} J$

So: $Energy \; in \; eV = \frac{3.97 \times 10^{-19}}{1.6 \times 10^{-19}} = 2.481 \, eV$

A20. Consider an X-ray source. What will be the thickness of absorber required for the radiation level to be reduced to 1/6 of its initial value in terms of α if $\alpha = 0.5$ m^{-1} (3 decimal places)?

Using the equation: $I_x = I_o e^{-\alpha x}$ and rearranging:

$$\frac{I_x}{I_o} = e^{-\alpha x}$$

$$\frac{1}{6} = e^{-\alpha x}$$

Taking inverse logarithms:

$$-\frac{\ln\left(\frac{1}{6}\right)}{\alpha} = x$$

So for $\alpha = 0.5$ m^{-1}

$$x = -\frac{\ln\left(\frac{1}{6}\right)}{0.5} = 3.584 \text{ m}$$

A21. Use differentiation and the equation $I_x = I_o e^{-\alpha x}$ to show the rate of change of intensity with distance.

Using the equation $I_x = I_o e^{-\alpha x}$ and differentiating with respect to distance x:
$$\frac{dI_x}{dx} = (-\alpha) I_o e^{-\alpha x}$$ Hence:
$$\frac{dI_x}{dx} = -\alpha I_x$$

Using:
$$\frac{dI_x}{dx} = -\alpha I_x$$

$$\frac{1}{I_x}\frac{dI_x}{dx} = -\alpha$$

So:
$$\alpha = -\frac{1}{I_x}\frac{dI_x}{dx}$$

A22. If the rate of change of intensity at a distance x is $-6\ Wm^{-1}$ and the intensity at distance x is 3.7 W, what is the decay constant α in m^{-1} (2 decimal places)?

$$\alpha = -\frac{1}{I_x}\frac{dI_x}{dx}$$

and substituting for values given:

$$\alpha = \frac{1}{3.7}6 = 1.62\ m^{-1}$$

A23. If the individual photon packet of an X-ray has a frequency of 1.7 ×10¹⁹ Hz and the overall intensity of the X-rays is recorded as 1.5 ×10⁹ Wm⁻² across an area of 1 square metre in 1 second what will be the total number of X-ray photons produced with constant flux?(2 significant figures).

Energy in an individual X-ray photon

$$E = hf = 6.62 \times 10^{-34} \times 1.7 \times 10^{19}$$
$$= 2.8254 \times 10^{-15} \, J$$

$$\text{Total number of X-ray photons} = \frac{\text{Total Energy}}{\text{Energy in an individual X-ray photon}}$$
$$= \frac{1.5 \times 10^9}{2.8254 \times 10^{-15}} = 5.3 \times 10^{23}$$

X-ray photons.

A24. Find the critical angle for a light wave travelling from diamond (refractive index = 2.4), into water having refractive index = 1.333 (2 decimal places).

$N_i \sin \vartheta_i = N_t \sin \vartheta_t$

$N_1 \sin \vartheta_c = N_2 \sin 90$

$\sin \theta c = \dfrac{N_2}{N_1} = \dfrac{1.333}{2.4} = 0.5554167$

So $\theta c = \sin^{-1}\left(\dfrac{N_2}{N_1}\right) = \sin^{-1} 0.5554167$

$= 33.74 \text{ degrees}$

A25. Use the following equation to calculate the missing variable in each situation:

λ = dx/L where λ is the wavelength of the light, x is the fringe spacing, d is the distance between slits, and L is the distance from the slits to the screen.

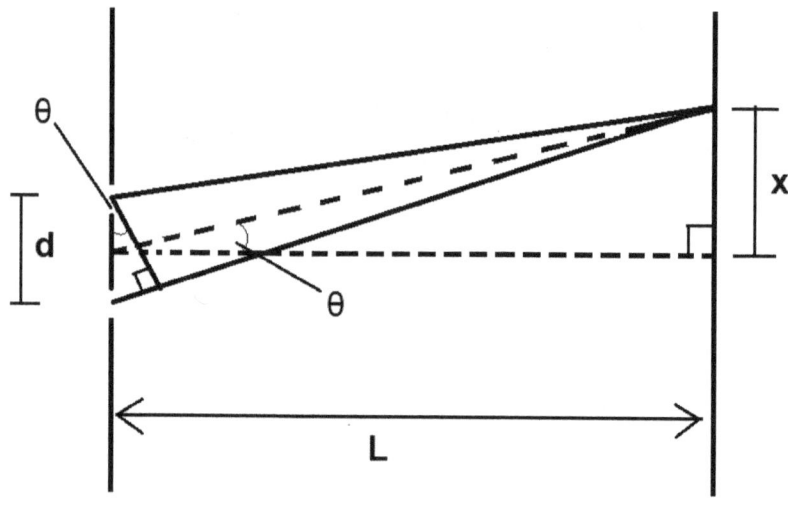

i) The wavelength of the light in the instance where x = 0.04 cm, d = 2.5 mm and L = 2.3 m (nearest whole nm).

ii) What distance must the screen be from the slits for the fringe spacing to be twice the distance between the slits for violet light when x = 1 mm?

i) Using $\lambda = dx/L$ and substituting:

$\lambda = 2.5 \times 10^{-3} \times 0.04 \times 10^{-2} / 2.3 = 434.78$ nm

ii) Rearranging $L = dx/\lambda$

and substituting: $= 0.5 \times 10^{-3} \times 1 \times 10^{-3} / 400 \times 10^{-9} = 1.25$ m

A26. Consider a wave incident at a boundary between air and water. The air has a refractive index of 1.0003, whilst the water has a refractive index of 1.333. The wave is incident at an angle of 60 degrees to the normal. Find the transmitted angle θ_t in the water, and sin θ_t (both 3 decimal places).

Using: $N_i \sin \vartheta_i = N_t \sin \vartheta_t$

$1.0003 \sin 60 = 1.333 \sin \vartheta_t$

Thus $\sin \vartheta_t = 0.6498764$ and $\vartheta_t = 40.532$ degrees.

A27. Consider a wave incident at a boundary between water and glass. The air has a refractive index of 1.333, whilst the glass has a refractive index of 1.52. The wave is incident at an angle of 10 degrees to the normal. Find the transmitted angle θ_t in the water, and $\sin \theta_t$ (both 3 decimal places).

Using: $N_i \sin \vartheta_i = N_t \sin \vartheta_t$

$1.333 \sin 10 = 1.52 \sin \vartheta_t$

Thus $\sin \vartheta_t = 0.152$ and $\vartheta_t = 8.759$ degrees

A28. A radar transmitter has a 15 W power source, if the source radiates isotropically in all directions what will the intensity be 20 m from the source (3 significant figures)?

Using the equation: $I = \dfrac{P}{4\pi R^2}$

Substituting: $I = \dfrac{15}{4\pi 20^2} = 2.98 \times 10^{-3}\ Wm^{-2}$

A29. A radar set detects a target at a range of 55 km. The intensity of the signal striking the target is 13×10^{-2} W m^{-2}.

Calculate the new intensity if the range is decreased to 25 km and the transmitted power is increased by 40 % (3 decimal places)?

$I_1 = \dfrac{P_1}{4\pi R_1^2}$ (29.1) and $I_2 = \dfrac{P_2}{4\pi R_2^2}$ (29.2)

Dividing equation 3.1 by equation 3.2 gives:

$$\frac{I_2}{I_1} = \frac{P_2}{P_1} \times \frac{R_1^2}{R_2^2} \quad (29.3)$$

Substituting for the values given:

$$\frac{I_2}{I_1} = 13 \times 10^{-2} \times \frac{1.4P}{P} \times \frac{55 \ km^2}{25 \ km^2}$$

$$I_2 = 13 \times 10^{-2} \times 1.4 \times \frac{55}{25}$$

$$I_2 = 0.088 \ W \ m^{-2}$$

A30. For a target at a range of 20 km and the same target then recorded at a range of 65 km what will be the ratio of the detected echo strength, (4 decimal places)?

For echoes:

$$\frac{I_2}{I_1} = \frac{P_2}{P_1} \times \frac{R_1^4}{R_2^4}$$

$$\frac{I_2}{I_1} = \frac{P}{P} \times \frac{20^4}{65^4} = \frac{20^4}{65^4} = 8.9633 \times 10^{-3}$$

A31. A radar set can just detect a target at a maximum range of 100 km. If the power output of the transmitter is increased by 35% what will be the new maximum range of the radar in km (4 significant figures)?

For echoes: $\dfrac{I_2}{I_1} = \dfrac{P_2}{P_1} \times \dfrac{R_1^4}{R_2^4}$ At the maximum detection range the receiver will receive the minimum intensity I_{min} for that receiver.

$$\dfrac{I_{min}}{I_{min}} = \dfrac{1.35P}{P} \times \dfrac{100^4}{R_2^4}$$

$$1 = 1.35 \times \dfrac{100^4}{R_2^4}$$

$$R_2^4 = 1.35 \times 100^4$$

R_2 = 107.8 km

A32. What is the ratio of scattering intensity for a wavelength of 550 nm compared with a wavelength of 750 nm (4 decimal places)?

$I_s \propto \frac{1}{\lambda^4}$ Then: $I_{550nm} \propto \frac{1}{550^4}$ and $I_{750nm} \propto \frac{1}{750^4}$

Dividing these two inequalities:

$\frac{I_{550nm}}{I_{750nm}} = \frac{750^4}{550^4} = 1.3636364^4 = 4.578$

A33. Using the beam attenuation coefficient $C(\lambda)$ at 575 nm and for a water depth of 10 m and incident light intensity I_{in} = 4 Wm^{-2} find the outgoing light intensity I_{out} (2 decimal places).

Use the equation: $I_{out} = I_{in}\, e^{-C(\lambda)x}$.

$C(\lambda) = 0.096$ m^{-1} so

$I_{out} = I_{in}\, e^{-C(\lambda)x} = 4 \times e^{-0.096 \times 10} = 1.53$ Wm^{-2}

A34. For water at a 470 nm wavelength, with a water depth of 7 m, with $C(\lambda) = 0.02$ m^{-1}

and surface incident light intensity $I_{in} = 2.6$ Wm^{-2}, what is the light intensity I_{out}

recorded at the sea bed (2 decimal places)?

$$I_{out} = I_{in}\, e^{-C(\lambda)x} = 2.6 \times e^{-0.02 \times 7} = 2.26 \text{ Wm}^{-2}$$

A35. The vertical and horizontal beam widths of a radar antenna are 60° and 2° respectively. The radar operates at 3.0 GHz. Calculate the vertical and vertical horizontal dimensions of the antenna (2 decimal places).

$$D_{horizontal} = \frac{60\lambda}{\alpha_{horizontal}} = \frac{60c}{f\alpha_{horizontal}} = \frac{60 \times 3 \times 10^8}{3.0 \times 10^9 \times 2} = 3.00 \text{ m}$$

$$\alpha_{vertical} = \frac{60\lambda}{D_{vertical}} = \frac{60c}{f \times D_{vertical}}$$

$$= \frac{60 \times 3 \times 10^8}{3.0 \times 10^9 \times 60} = 0.10 \text{ m}$$

A36. The vertical and horizontal dimensions of a radar antenna are 15 cm and 5.2 m respectively. If the radar operates at 3.2 GHz, what are the vertical and horizontal beam widths (2 decimal places)?

$$\alpha_{horizontal} = \frac{60\lambda}{D_{horizontal}} = \frac{60c}{f \times D_{horizontal}} = \frac{60 \times 3 \times 10^8}{3.2 \times 10^9 \times 5.2} = 1.08^o$$

$$\alpha_{vertical} = \frac{60\lambda}{D_{vertical}} = \frac{60c}{f \times D_{vertical}} = \frac{60 \times 3 \times 10^8}{3.2 \times 10^9 \times 0.15} = 37.5^o$$

A37. A diffraction grating of 4000 lines per centimetre is used with white light. How many orders of spectra will be observed?

No. of lines per metre = 4×10^5

Spacing d = 2.5×10^{-6}

For the *nth* constructive order for a grating
$$d \sin \theta = n\lambda$$
The maximum value of θ incident is 90°, so sin (90°) = 1

Taking the spectral range for white light to be between 400 nm and 700 nm the mid-point of the visible spectrum observed will be about 550 nm or 550×10^{-9} m

So the *maximum* possible value of n is:

$$n = \frac{d \sin \theta}{\lambda} = \frac{2.5 \times 10^{-6} \times \sin 90}{550 \times 10^{-9}}$$
$$= 4.545 \text{ (2 decimal places)}$$

Therefore 4 complete orders should be seen.

A38. With two receivers calculate the resolution limit if the wavelength is 2 m and the two receivers are spaced 2 km apart (2 decimal places).

The angular separation = $60\lambda/D = 60 \times \dfrac{2}{2 \times 10^3} = 0.06$ degrees.

A39. With two receivers calculate the resolution limit if the wavelength is 1 m and there are now 30 receivers spaced 4 km apart (4 significant figures).

The array angular separation = $60\lambda/D = 60\lambda/(n-1)d$ = $60 \times \dfrac{1}{(30-1) \times 1 \times 10^3} = 2.069 \times 10^{-7}$ degrees.

A40. What is the receiver spacing in metres if a linear array can resolve a beam width of $0.02°$? The wavelength is 2 cm with 30 elements in a *sonar* array (2 decimal places).

Using the equation: the array angular separation $0.02°$
$= 60\lambda/(n-1)d = 60\times \dfrac{0.02}{(30-1)\times d}$

Rearranging for d: $d = 60\times \dfrac{0.02}{(30-1)\times 0.02} = 2.07$ m apart.

A41. If a receiver R a distance x from both sources A and B is moved a distance *dx* towards source A calculate the distance need to move from the central maximum to the *second* destructive minimum.

$$\frac{dx}{\lambda} = \frac{\phi}{360}$$

```
              R
    x- dx     ▣         x + dx
              dx
■━━━━━━━━━━━━━━━━━━━━━━━━━━━■
A                            B
```

Path Difference d at C = (x-dx) − (x-dx) = 2dx

Thus a half wavelength path difference is equivalent to a physical movement of one quarter wavelength at the first destructive minima, i.e.

$$dx = \frac{\lambda}{4}$$ which is the first out of phase condition.

Hence $dx = \frac{\lambda}{4} + \frac{\lambda}{2} = \frac{3\lambda}{4}$ gives rise to a phase equivalent difference of $dx = \frac{3\lambda}{2}$ the second out of phase condition.

A42. For two waves of the same frequency emitting with a time delay of T/3 and a physical path distance of $\lambda/4$ and a phase difference of 180 degrees what will be the resultant of the two waves arriving at the same point in space?

Using the representation: $\frac{t}{T} = \frac{d}{\lambda} = \frac{\phi}{360}$

For the path difference: $\frac{t}{T} = \frac{d}{\lambda}$ so $t = \frac{dT}{\lambda}$ and,

The equivalent time delay for the path difference = $\frac{\lambda/4 \, T}{\lambda} = T/4$ and,

The equivalent time delay for the phase difference $\frac{t}{T} = \frac{\phi}{360}$ so $t = \frac{\phi T}{360} = \frac{180T}{360} = T/2$

and considering the time delay, path difference, and phase difference in terms of equivalent time respectively then:

The total time delay = T/3 + T/4 + T/2 = 13T/12

A43. Find the time delay required in terms of the Periodic time T if the path difference between two adjacent sources is 3 cm and the wavelength is 5 cm?

The time delay equivalence can be found from the expression: $\frac{t}{T} = \frac{d}{\lambda}$

the time delay required if the path difference between two adjacent sources is 3 cm will be $t = \frac{dT}{\lambda} = \frac{0.03\,T}{0.05} = 0.6T$.

A44. Consider a regularly spaced square array of sources in both the vertical and horizontal direction. If there are 1600 sources in total and the sources in every row and column are spaced by half a wavelength what will be the minimum resulting beam width if the wavelength is 1 cm (3 decimal places)?

The total number of sources is the square of the number of sources along any row or column, i.e. the number of sources n in any row or column is related to the total number of sources N in a square array by the equation: $N = n^2$

So $n = \sqrt{N}$
Hence $n = \sqrt{1600} = 40$

Now the number of *spaces* is related to the number of sources by the equation:

$$number\ of\ spaces = number\ of\ sources - 1$$

So the number of spaces = 40 -1 = 39

Each space d has a value of $\frac{\lambda}{2} = \frac{1}{2} = 0.5$ cm (but actual wavelength value is not required see below).

The beam width equation gives:

$$\alpha = \frac{60\lambda}{D} = \frac{60\lambda}{(n-1) \times d}$$

$$= \frac{60\lambda}{39 \times \frac{\lambda}{2}} = \frac{60}{19.5} = 3.077 \text{ degrees.}$$

A45. Find the time delay required to steer a radar beam off the bore sight (the normal straight through direction) if the required path difference between adjacent sources is 3 cm.

Given that speed = path difference/ time delay $c = \frac{path\ difference}{time\ delay}$

Therefore: $time\ delay = \frac{0.03}{3 \times 10^8} = 1.0 \times 10^{-10}$ seconds.

A46. If the transmitted radar frequency is 3 GHz and the relative velocity is 20 metres per second and the speed of light c is taken as usual find the one way frequency shift observed (1 decimal place).

Since: $\Delta f = \frac{V_{relative}}{V} f_{transmitted}$

$$\Delta f = \frac{20}{3 \times 10^8} \times 3 \times 10^9 = 200 \; Hertz$$

A47. In the case of a Doppler shifted echo, calculate the velocity of the target if the observer is stationary, the transmitted frequency is 8.9 GHz and the echo shift is 4 kHz (3 significant figures).

For echoes: $\Delta f \; echoes = \frac{2V_{relative}}{c} f_{transmitted}$

So rearranging: $V_{relative} = \frac{c \times \Delta f \; echoes}{2 f_{transmitted}}$

And substituting: $V_{relative} = \frac{3 \times 10^8 \times 4 \times 10^3}{2 \times 8.9 \times 10^9} = 67.4$

metres per second.

A48. $V_{relative} = V_{source} \cos \theta_{source} - V_{target} \cos \theta_{target}$ to calculate the echo Doppler shift if the source has a value of 50 metres per second on a bearing of 60 degrees and the target has a value of 15 metres per second on a bearing of 130 degrees and the radar is transmitting on a frequency of 9.5 GHz.

Use the following expression for relative velocity:
$$V_{relative} = V_{source} \cos \theta_{source} - V_{target} \cos \theta_{target}$$

$$V_{relative} = 50 \cos 60 - 15 \cos 130$$
$$= 34.64 \text{ metres per second}$$

And using $\Delta f \text{ echoes} = \dfrac{2V_{relative}}{c} f_{transmitted}$: and substituting:

$$\Delta f \text{ echoes} = \dfrac{2 \times 34.64}{3 \times 10^8} \times 9.5 \times 10^9 = 2193.98 \text{ Hz}$$

A49. An amplifier has a power gain of + 15 dB. If an input power is 13 mW dB, what is the output signal power (3 significant figures)?

Using the Power ratio expression in decibels

$= 10 \log (P_2/ P_1)$

So: $+15 = 10 \log (P_2/ 0.013)$ and taking antilogs $P_2 = 0.013 \times 10^{1.5} = 0.411$ W

A50. If the initial power level P_1 is 2W and the final power level P_2 is 110W what is the Power ratio in decibels (2 decimal places)?

decibels $= 10 \log (P_2/ P_1)$

So decibels $= 10 \log (110/ 2) = 17.40$ dB

LIST OF EQUATIONS AS USED AND KEY CONSTANTS

$V = f\lambda$ V is speed, $=f$ is frequency and λ is wavelength.

$\dfrac{t}{T} = \dfrac{\phi}{360}$ t is time delay, T the period, and ϕ is the phase difference between two waves.

$$\text{speed v} = \dfrac{distance\ d}{time\ t\ taken\ to\ cover\ this\ distance}$$

$f = \dfrac{1}{T}$

Group velocity $V_g = \dfrac{\omega}{k}$ where ω is the angular velocity and k the wave vector.

Phase velocity $V_p = \frac{d\omega}{dk}$ where $d\omega$ is the change in angular velocity and dk is the change in the wave vector.

$y(t) = \cos(\omega t)$ Repetitive wave motion displacement in the y direction as a function of time t.

For a pendulum $\omega = \sqrt{\frac{g}{l}}$ where g is the acceleration due to gravity 9.81 metres per second per second, and l is the length of the pendulum.

For waves in deep water speed, $c = \sqrt{\frac{g\lambda}{2\pi}}$

For waves in shallow water, $c = \sqrt{gz}$

Deep water wave amplitude $A = A_0 \exp\left(-\frac{2\pi z}{\lambda}\right)$ A_0 is the initial wave amplitude, A the amplitude as a function of depth z and wavelength λ.

$$\frac{p}{\rho} + \frac{v^2}{2} + gH = constant$$

Bernouilli's equation relating pressure p, ρ density, velocity v, gravitational constant g and H the Height of the water column.

$E = hf$ E is the energy of a photon is determined by Planck's constant h = 6.626176 x 10^{-34} Js and the frequency f of the wave.

Energy of a photon from X Joules to electron volts is given by:

$$Energy\ in\ eV = \frac{X \times 10^{-19}}{1.6 \times 10^{-19}}$$

where $1\ eV = 1.6 \times 10^{-19} J$

$I_x = I_o e^{-\alpha x}$ Intensity reduction through a thickness x of absorber with attenuation coefficient α in m^{-1}

$N_i \sin \vartheta_i = N_t \sin \vartheta_t$ Relates the incident refractive index N_i and the incident angle ϑ_i to the transmitted refractive index N_t and the transmitted angle ϑ_t.

$\lambda = dx/L$ where λ is the wavelength of the light, x is the fringe spacing, d is the distance between slits, and L is the distance from the slits to the screen.

$$\frac{I_2}{I_1} = \frac{P_2}{P_1} \times \frac{R_1^2}{R_2^2}$$

One way spreading intensity equation for the change in intensity due to changes in both power output P and range R.

$$\frac{I_2}{I_1} = \frac{P_2}{P_1} \times \frac{R_1^4}{R_2^4}$$

Echo changes in intensity due to changes in both Power P and range R.

$I_s \propto \frac{1}{\lambda^4}$ relationship between scattering intensity I and wavelength λ.

$I_{out} = I_{in}\, e^{-C(\lambda)x}$ Intensity output dependent upon initial intensity, distance x and the attenuation coefficient $C(\lambda)$ as a function of wavelength.

$$\alpha_{vertical} = \frac{60\lambda}{D_{vertical}} = \frac{60c}{f \times D_{vertical}}$$

Beam width depends upon: wavelength, aperture width, frequency f and speed of electromagnetic waves c = 3×10^8 ms^{-1}.

For the *nth* constructive order for a grating

$d \sin\theta = n\lambda$ where d is the grating spacing θ the angle of incidence, and λ is the wavelength.

PAGES FOR WORKING AND CALCULATIONS

OTHER BOOKS BY CHRISTOPHER LAVERS

Crosstalk- A reflection in Faith Christian poetry, October 2011 ISBN: 978-1-4709-0122-6. (Publisher: Lulu Enterprises, Inc., Rayleigh, North Carolina).

Stealth Warship Technology (ISBN 9781408175255), in the Reeds Marine Engineering and Technology series, *Adlard Cole*, Published October 2012.

Basic Electromagnetic Wave Concepts For Engineers ISBN 978-1-4709-5404-8. (Publisher: Lulu Enterprises, Inc., Rayleigh, North Carolina).

Basic Electrotechnology Reeds Vol 6: (Reed's Marine Engineering) Publisher: Adlard Coles ISBN: 0713668385 DDC: 797 Edition: Paperback; 2008-01-01 Publication date: December 2012.

Recent Developments in Remote Sensing for Human Disaster Management and Mitigation- Natural and Man-made 2013 Editor Christopher Lavers 978-1-291-22463-4 (Publisher: Lulu Enterprises, Inc., Rayleigh, North Carolina), December 2012.

EXtreme Faith Walking the Talk Various motivational Christian authors, edited by Christopher Lavers, March 2013 ISBN: 9781291351088 (Publisher: Lulu Enterprises, Inc., Rayleigh, North Carolina).

Advanced Electrotechnology Reeds Vol 7: (Reed's Marine Engineering) Publisher: Adlard Coles ISBN: 0713676841 DDC: 621.30246238 Edition: Paperback; Publication date: April 2014.

Introductions: Physics Wave Concepts for Marine Engineering Applications by Christopher Lavers and Sara-Kate Lavers, March 2017, ISBN: 9781472922151 Adlard Cole Nautical.

Introductions: Essential Sensing and Telecommunications for Marine Engineering Applications by Christopher Lavers, March 2017, ISBN: 97814792922182 Adlard Cole Nautical.

Further Physics Wave Concepts for Marine Engineering Applications (in press Lulu.com)

FICTION *TALES FROM THE FOREST SERIES:*

 # 1 *ON THE WINDS OF THE DESERT SANDS* *November 2012, ISBN: 978-1-291-08012-4* Publisher: Lulu Enterprises, Inc., Rayleigh, North Carolina), 2012.

 # 2 *ESCAPE FROM WARSAW* *To be printed 2017.*

 # 3 *TALES FROM THE WESTERN JUNGLE.*

www.ingramcontent.com/pod-product-compliance
Lightning Source LLC
Chambersburg PA
CBHW070426180526
45158CB00017B/773